Peter Puchala

Wilderness as a possibility for national parks in Slovakia

Peter Puchala

Wilderness as a possibility for national parks in Slovakia

Feasibility check for Tatra National Park's Potencial for PAN Park certification

LAP LAMBERT Academic Publishing

Impressum / Imprint
Bibliografische Information der Deutschen Nationalbibliothek: Die Deutsche Nationalbibliothek verzeichnet diese Publikation in der Deutschen Nationalbibliografie; detaillierte bibliografische Daten sind im Internet über http://dnb.d-nb.de abrufbar.
Alle in diesem Buch genannten Marken und Produktnamen unterliegen warenzeichen-, marken- oder patentrechtlichem Schutz bzw. sind Warenzeichen oder eingetragene Warenzeichen der jeweiligen Inhaber. Die Wiedergabe von Marken, Produktnamen, Gebrauchsnamen, Handelsnamen, Warenbezeichnungen u.s.w. in diesem Werk berechtigt auch ohne besondere Kennzeichnung nicht zu der Annahme, dass solche Namen im Sinne der Warenzeichen- und Markenschutzgesetzgebung als frei zu betrachten wären und daher von jedermann benutzt werden dürften.

Bibliographic information published by the Deutsche Nationalbibliothek: The Deutsche Nationalbibliothek lists this publication in the Deutsche Nationalbibliografie; detailed bibliographic data are available in the Internet at http://dnb.d-nb.de.
Any brand names and product names mentioned in this book are subject to trademark, brand or patent protection and are trademarks or registered trademarks of their respective holders. The use of brand names, product names, common names, trade names, product descriptions etc. even without a particular marking in this works is in no way to be construed to mean that such names may be regarded as unrestricted in respect of trademark and brand protection legislation and could thus be used by anyone.

Coverbild / Cover image: www.ingimage.com

Verlag / Publisher:
LAP LAMBERT Academic Publishing
ist ein Imprint der / is a trademark of
OmniScriptum GmbH & Co. KG
Heinrich-Böcking-Str. 6-8, 66121 Saarbrücken, Deutschland / Germany
Email: info@lap-publishing.com

Herstellung: siehe letzte Seite /
Printed at: see last page
ISBN: 978-3-659-51862-1

Wilderness as a possibility for national parks in Slovakia

Feasibility check of the Tatra National Park's potential for PAN Parks certification

Author: Peter Puchala

Table of Contents

Acknowledments 3

1   **Introduction** 4

   1.1   Protected areas and wilderness 4
   1.2   PAN Parks 8

2   **Project Description** 13

   2.1   Situation with PAN Parks and wilderness in Slovakia 13
   2.2   Methodology 15
   2.3   Objectives and purposes 15

3   **Results** 17

   3.1   General overview of Tatra National Park 17
   3.2   Natural values 18
   3.3   Habitat management 22
   3.4   Visitor management 29
   3.5   Assessment of Tatra National Park potential 32
   3.6   SWOT analysis from wilderness management point of view 37

4   **Discussion and recommendations** 40

5   **Conclusion** 46

6   **References** 48

   6.1   Literature 48
   6.2   Internet Resources 50

Acknowledgments

This book represents a final step of my studies during the years 2009 - 2011 in the MSc programme "Management of Protected Areas" at Klagenfurt University. I am very grateful to everybody who contributed to the results of this work, encouraged me and supported me in different ways during two years of studies. I would like to thank all the people who provided me with any data and information about the topic and namely to Slavomír Celer from Tatra National Park administration with whom I communicated most of time. Then I would like to thank Robert Rajchl who was willing to provide me some photographs from Tichá and Kôprová valleys in Tatra National Park and Ben Sweeney for proof reading.

In particular I would like to thank Zoltan Kun from PAN Parks Foundation who was willing to be my supervisor. I highly appreciate his help and support in the ideas of the study, all recommendations, advices and comments that helped me to improve my work.

Last but not least I would like to thank all my classmates and lecturers for inspiring and motivating and providing a friendly atmosphere during the whole two years studies in Klagenfurt.

1 Introduction

1.1 Protected areas and wilderness

Protected areas are the fundamental building blocks of virtually all national and international conservation strategies, supported by governments and international institutions such as the Convention on Biological Diversity (CBD). They provide the core of efforts to protect the world's threatened and rare species and are increasingly recognized as essential providers of ecosystem services and biological resources; key components in climate change mitigation strategies; and in some cases also vehicles for conservation of threatened human communities or sites of great cultural and spiritual value (DUDLEY 2008).

The global number and extent of nationally designated protected areas has increased dramatically over the past century. By 2008, there were over 120,000 protected areas covering a total of about 21 million square kilometers of land and sea. While the terrestrial PA's listed in the World database on Protected Areas (WDPA) cover 12.2 % of the Earth's land area, marine PA's currently cover 5.9 % of the Earth's territorial seas and only 0.5 % of the extraterritorial seas (www.wdpa.org).

IUCN has developed a standard international classification system of six categories of PA's. This system has been developed under IUCN Guidelines for Protected Area Management Categories and forms the organizational structure of UN List of protected areas. Definitions of the IUCN Protected Area Management Categories are described below:

- Category Ia: Strict Nature Reserve: protected area managed mainly for science
- Category Ib.: Wilderness Area: protected area managed mainly for wilderness protection
- Category II: National Park: protected area managed mainly for ecosystem protection and recreation (natural area of land and /or sea)
- Category III: Natural Monument: protected area managed mainly for conservation of specific natural features
- Category IV.: Habitat/Species Management Area: protected area managed mainly for conservation through management intervention
- Category V.: Protected Landscape/Seascape: protected area managed mainly for landscape/seascape conservation and recreation
- Category VI.: Managed resources Protected Area: protected area managed mainly for the sustainable use of natural ecosystems

The management objectives for all above mentioned IUCN categories are described in Tab. 1.

Management objectives	Management categories						
	Ia	Ib	II	III	IV	V	VI
Scientific research	1	3	2	2	2	2	3
Wilderness protection	2	1	2	3	3	-	2
Species conservation	1	2	1	1	1	2	1
Maintenance of environmental services	2	1	1	-	1	2	1
Protection of specific natural/cultural features	-	-	2	1	3	1	3
Tourism and recreation	-	2	1	1	3	1	3
Education	-	-	2	2	2	2	3
Sustainable use of resources from natural ecosystems	-	3	3	-	2	2	1
Maintenance of cultural/traditional attributes	-	-	-	-	-	1	2

Tab. 1: Matrix of management objectives and IUCN protected area management categories

Legend: 1 – primary objective, 2 – secondary objective, 3 – potentially applicable objective – not applicable (according to Dudley 2008)

Interpretation of wilderness is possible under the umbrella of the IUCN category system. Within all six of the IUCN Categories the definition is best covered by category Ib – Wilderness Area. However, the objectives of the next two categories Ia –Strict Nature Reserve and II – National Park are also in line with wilderness management objectives.

It is often cited that wilderness areas currently occupy between 1 % and 2 % of Europe's land surface; however, it was proposed that for the calculation of total wilderness areas, category Ia should be added to category Ib Protected Areas as well, containing the intact and functioning natural habitats and processes that characterize wilderness but not labeled as such nor opened to public access; further hitherto unrecorded areas of wilderness are also likely to lie within the boundaries of category II of the IUCN categorization (COLEMAN AND AYKROYD 2009).

One of the possible definitions of wilderness area is one such as the IUCN category Ib that defines a wilderness area as a large unmodified or slightly modified area, retaining its natural character and influence, without permanent of significant human habitation, which is protected and managed so as to preserve its natural condition (Dudley 2008). Wilderness areas represent a vital element of natural and cultural heritage and provide important economic, social and environmental benefits, including ecosystem services for local communities, land owners and society at large.

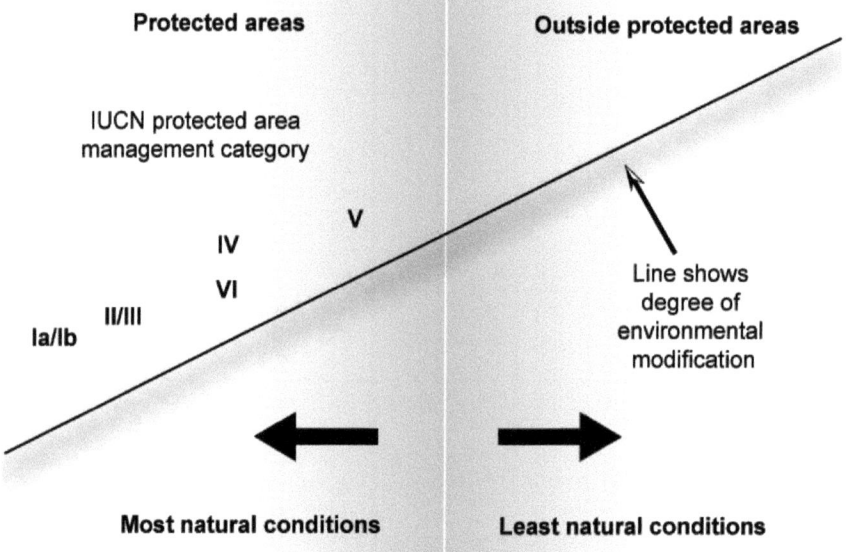

Fig. 1: Naturalness and IUCN protected area categories (according to Dudley 2008)

Wilderness areas should fulfill several criteria. They can be usefully divided into three zones with a core area surrounded by a buffer area of minimal activities that leads to a transition zone. Currently there is a discussion between several NGO's about the definition of wilderness[1]. One of the most important criteria is that of size. According to this discussion the core area should have minimum size of 3,000 ha with an aspiration of 10,000 ha to be achieved if possible within a stated timescale and the buffer zone should be large enough to allow expansion of the core zone. However, some countries have their own definition of the basic characteristics of wilderness areas (e.g. in Finland a wilderness area should comprise a minimum of 15,000 ha and usually be more than 10 km in width (SAARINEN 1998)). Other criteria of wilderness areas are biodiversity, natural processes, no permanent settlements, infrastructure and roads, no forestry, hunting and fishing in the core zone. There should be no natural resource use including berry, mushroom and nut picking, and grazing livestock, except by special agreements for subsistence of indigenous people. Wilderness areas should be free to the public, though with management to minimize the impact, and some sport and recreational activities with some spatial restrictions could be possible.

---

[1] Discussion between several NGO's forming so-called Wilderness Working Group

Wilderness is best understood as a multidimensional concept, consisting of biological and social elements. In both a global and European context, the term is used essentially as a biological descriptor, bearing no indication to an area's status of being protected or having specific social and legal characteristics. Generally speaking, wilderness areas can be described as large territories without major human interference, the lack of which enables natural processes to occur and wildlife to live in their natural ecological status and natural habitats (PAN Parks 2009).

Wilderness and wild areas are important because of their indirect and direct economic, health, social, research, and cultural values. They have a high intrinsic value, are essential laboratories for research of biological diversity, natural and ecological processes and they provide genetic banks for the future. They can also contribute to adaptation and mitigation to climate change and provide a wide range of ecosystem services. Moreover, they are an important part of the strategy for halting biodiversity loss and promoting natural ecosystem processes and functions (COLEMAN AND AYKROYD 2009).

The importance of wilderness areas is undoubtedly very high and they usually provide several benefits and ecological values. The most important ecological values are as follows (PAN Parks 2009):

- a refuge for endangered species and a home of undiscovered species
- habitats with highly adapted fauna and flora, which depend on the existence of these areas;
- reference laboratories where the natural processes of evolution still continue;
- restoration of natural dynamics after natural disturbances.

Wilderness also offers strong, sustainable economic, social, cultural and spiritual benefits like:

- opportunities for nature-based ecotourism supporting local and rural development;
- places for inspiration, renewal and recreation far from the cities and pressures of modern life;
- the potential to help tackle important city issues such as youth development and health care;
- addressing climate change through several ecological services such as carbon sequestration, flood mitigation and others.

Another importance of wilderness is that its protection can help maintain opportunities to continue a traditional relationship with nature. In this meaning we can use also the term "traditional ecological knowledge", which forms part of these relationships and has been acknowledged as a contributor to understanding the effects of management decisions and human-use impacts on long-term ecological composition, structure, and function (WATSON et al. 2003).

Recently, in the last few years the wilderness and its conservation has become a very important issue of the European Nature Conservation Strategy. The European Commission and European Parliament have adopted documents which ensure biodiversity and wilderness conservation. European Commission has recently announced the EU Biodiversity Strategy to 2020 where the headline target is halting the loss of biodiversity and degradation services in the EU by 2020 and restoring them so far as feasible. The conservation of wilderness areas is one of the important measures for achieving this target in forestry management[2].

Another very important document for wilderness conservation policy is the European Parliament Resolution on Wilderness in Europe[3]. One of the parts of this document calls on commission and the member states to develop wilderness areas, and stresses the need for provision of special funding for reducing fragmentation, careful management of re-wilding areas, development of compensation mechanisms and programmes, raising awareness, building understanding and introducing wilderness-related concepts such as the role of free natural processes into the monitoring and measurement of a favourable conservation status and consider that this should be done in cooperation with the local population and other relevant stakeholders.

There has also been quite a growing interest in wilderness and wild areas among nature conservationist in the recent period. The latter is evident from the relatively recent establishment of a number of European-based organizations and initiatives dedicated to increasing the profile and promotion of the wilderness concept and philosophy. The most long-established among these is PAN Parks, but more recently the Wild Europe Initiative, which is supported among others by the EUROPARC Federation, and the Wilderness Foundation and the Large Herbivore Foundation have become active in the field (JONES-WALTERS & ČIVIĆ 2010).

1.2 PAN Parks

PAN Parks (PAN – Protected Area Network) is a European network of the best managed protected areas. The aim of this network is to improve management of protected areas and nature conservation using sustainable tourism as a tool. PAN Parks create a network of protected areas certified in accord with PAN Parks quality standards, which cover relevant wilderness protection, social, economic and cultural aspects. The core idea of PAN Parks includes large natural areas of

---

[2] EU Biodiversity Strategy
    (http://ec.europa.eu/environment/nature/biodiversity/comm2006/2020.htm)
[3] European Parliament resolution of 3 February 2009 on Wilderness in Europe

wilderness which welcome visitors and provide them with an outstanding access to wildlife. However, nature conservation highly prevails over managerial activities in order to achieve an effective habitat protection with the lowest environmental impact.

The starting point in the PAN Parks project is wilderness protection. Nowadays nature conservation organizations are increasingly starting to realize that socio-cultural and economic sustainability in a region with a (protected) natural area are equally important when it comes to nature preservation. Tourism could be an instrument in sustainable development and nature conservation, giving nature an economic value (and as such preserve it) and at the same time it benefits socio-cultural sustainability (BEUNDERS 2002).

The concept of PAN Parks was founded in the year 1997 as a cooperative initiative of the WWF and Molecaten group, which is a leisure company from Netherlands providing support in the field of sustainable tourism. The aim of this partnership was to find a balance between nature, parks and human activities. The Certified PAN Parks label sets high standards for the management of protected areas from both a conservation and a tourism point of view. By developing common communications and a marketing platform of partner areas, the Foundation wants to raise awareness of the European protected areas in general. There are four main factors that make PAN Parks different from other verification systems (Engeldorp 2002):

- Independent verification aiming in improving the quality of management and the wilderness experience for visitors in European protected areas;
- Strong marketing of the concept and Certified PAN Parks in cooperation with tourism business;
- Developing a common communications approach towards people interested in green tourism;
- Creating partnerships with local communities and local or small scale businesses in order to implement a sustainable tourism development strategy on a regional scale.

In order to protect the most seriously threatened habitats and species across Europe, NATURA 2000 network was established. The conservation of European wilderness, as one of the most effective tools in the protection of natural habitat types and species of Community interest is an integral part of this network. The PAN Parks initiative has been identified by DG Environment – the European Commission as one of the most relevant initiatives for managing tourism within NATURA 2000 sites (TAYLOR 2004).

The PAN Parks concept is a reliable trademark for tourism, recreation and nature. The candidate Parks have to undergo certification by an independent certifying organization under recognized standards. To guarantee constant quality, the certificate awarded to a park should be periodically reviewed. This review, or verification, is done based on principles, criteria and indicators (VAN DER DONK 2000).

The backbone of the foundation is a transparent certification process. If a protected area wants to become a certified PAN Park, it must meet each of PAN Park's principles and criteria. This process is aimed at defining the quality standards that both protected areas and local business partners must maintain in order to become and remain certified (Vančura et al., 2008).

Principles of PAN Parks

The PAN Parks principles and criteria were worked out by consultative WWF offices, several protected areas and several other NGO's (http://www.panparks.org/learn/apply-for-verification/principles-and-criteria). Principles and criteria can be regularly updated based on the field experience and usually when circumstances change. Principles and Criteria are key documents that clearly explain requirements for PAN Parks and are the core of the PAN Park approach. They also define the roles, rights and obligations of involved partners. The principles diverge into criteria and indicators and through their application ensure the nature conservation and a high quality, nature-based experience for visitors. They also ensure the sustainability of protected areas and its surrounding region.

There are five main principles for the certification of the protected area to become a PAN Park. The first three principles apply to the protected area and its management body, which is responsible for the management of the protected area. Principles 1 – 3 are meant to indicate when a protected area can be certified as a PAN Park (PAN Parks 2008). Principles four and five set criteria for a sustainable tourism development strategy and for participation of local partners. The five principles and what they stand for are described below:

Principle 1: Natural values
PAN Parks are large protected areas, representative of Europe's natural heritage and conserve international important wildlife and ecosystems.

Principle 2: Habitat management
Design and management of the PAN Park aims to maintain and, if necessary, restore the area's natural ecological processes and biodiversity.

Principle 3: Visitor management
Visitor management safeguards the natural values of the PAN Park and aims to provide visitors with a high-quality experience based on appreciation of nature.

Principle 4: Sustainable Tourism Development Strategy
The protected area administration and its relevant partners in the PAN Parks region aim at achieving a synergy between the conservation of natural values and sustainable tourism by jointly developing and implementing a sustainable tourism development strategy (STDS).

Principle 5: Business partners

PAN Parks' tourism-related business partners are legal enterprises that are committed to sustainable tourism and support the goals of certified PAN Parks. They actively cooperate with the Local PAN Park Group to implement the PAN Park region's sustainable tourism development strategy and ensure their business complies with high national and international standards of environment management.

The main aim of the first three principles is to ensure the conservation of natural values of a protected area through proper habitat and visitor management. Any protected area applying for PAN Parks certification has to define the scope of protection, the international importance, and the size of the protected area. One of the very important criteria in applying for certification is ensuring a sufficient size for the protected area. The size of a PAN Park area should cover at least 20,000 ha and the core zone of the area considered as a wilderness area should cover at least 10,000 ha. PAN Parks Foundation interprets wilderness along the lines of three major criteria. The protected area has an ecologically unfragmented wilderness areas of at least 10,000 hectares where no extractive uses are permitted and where the only management interventions are those aimed at maintaining or restoring natural ecological processes and ecological integrity (PAN Parks 2009).

Principle 4 states criteria for development of STDS. That is why this principle could be called the stakeholder principle. This strategy is a basic document for the planning of STDS and should be a result of the acceptance of a variety of interests of different stakeholder groups including local communities, municipalities, landowners and land users, and local businesses in tourism.

Principle 5 contains criteria and indicators for local business partners (especially for entrepreneurs in ecotourism) to fulfill STDS. This principle sets a minimum standard for them and should always have a local aspect relating business partners with the certified park and its surrounding region and a general aspect relating the business partners with the best possible environmental standards of the park's country (PAN Parks 2008).

Usually the whole process of PAN Parks verification takes place in three subsequent phases in which independent experts check the protected area according to all five of the PAN Parks principles. In the first phase the protected area is verified according to the first three principles. After fulfillment of this phase a sustainable tourism development strategy in accordance with principle 4 is verified and the third and final phase includes verification of the local business partners according to principle 5. Soomaa National Park in Estonia was the first national park that followed a different approach. During verification of this park all principles had to be fulfilled once. Since last year this approach is used in PAN Parks verification process.

PAN Parks within Europe

The PAN foundation carried out research in the year 2002 to define potential PAN Parks throughout the European countries based on selected PAN Parks indicators of principles and criteria (KUN 2002). Out of 2,926 researched protected areas 134 could be listed as potential PAN Parks. This list doesn't necessarily mean that all listed areas could fulfill all PAN Parks criteria but they have the potential. This study revealed big differences in PAN Park potentials among European countries. The results are shown in Fig. 2. Regions of the Atlantic coast, Western Europe and the western Mediterranean region have a very bad condition of natural processes in ecosystems. Ecosystem protection is in an alarming status in two regions: the Baltic and Adriatic coasts. Relatively good conditions and quite high potential for PAN Parks occur in regions of Eastern Mediterranean countries, central Europe, European parts of Russia and in Scandinavia.

The study mentioned above is almost ten years old and recent results of such a study could be different in some countries. Some countries are working more properly on the concept of wilderness. One of the examples is Germany, which was indicated in the study as a country with badly represented natural processes. The main management approach of this country to its national parks, which cover about 2.6 % of total area of Germany, is closely related to the wilderness concept (MARTIN ET AL. 2008). On the other hand some countries that were identified as ones with very high potential have currently practically no wilderness management, e.g. Hungary (Kun, pers. comm.). Due to changes and circumstances it would be very interesting to repeat this research and evaluate the current situation with a proportion of wilderness in all European countries and identify their PAN Parks potential.

During its 11-year existence PAN Parks established a network of protected areas that use the PAN Parks certification to ensure that their wilderness areas are safeguarded, enhanced and appreciated (PAN Parks). The following protected areas have been already certified as PAN Parks:

Archipelago National Park
Borjomi-Kharagauli National Park
Central Balkan National Park
Fulufjället National Park
Majella National Park
Oulanka National Park
Paanajärvi National Park
Peneda-Gerês National Park
Retezat National Park
Rila National Park
Soomaa National Park

Fig. 2: Country status of PAN Park potential

Legend: red: low potential for PAN Parks (natural processes are badly represented)

yellow: medium potential for PAN Parks/natural processes are in alarming situation

green: high potential for PAN Parks/natural processes are well-represented

stars indicate small European countries (Andorra, Liechtenstein, Malta, Monaco, San Marino and Vatican Holy State which were not researched)

2 PROJECT DESCRIPTION

2.1 Situation with PAN Parks and wilderness in Slovakia

Slovakia belongs to the European countries where some parts of wilderness can still be found. However, only 2 % of total coverage of the country belongs to protected areas with the highest degree of conservation where no interventions should be made.

The national system of protected areas in Slovakia consists of nine national parks and 14 protected landscape areas and a large amount of so called small scaled protected areas (national nature reserves, natural reserves, national nature monuments, nature monuments and protected ranges). The total coverage of protected areas within all these categories is approximately 23 % of territory of the country.

The Slovak Republic, as a member of European Union, has adopted habitat and bird directives of the European Commission into its national legislation when it entered the European Union in May 2004. A national list of proposed Special Protected Areas (SPAs) was ratified by resolution of the Government of the Slovak Republic and consists of 38 SPA's with a total area of 1,236,545 ha, which represents 25.2 % of the territory of Slovakia. A national list of proposed Sites of Community Importance (SCI's) was ratified by a resolution of the Government of the Slovak Republic and consists of 382 SCI's that totally covers 11.7 % of territory of Slovakia (www.sopsr.sk).

A study conducted by PAN Parks foundation in the year 2002 analyzed 39 areas in Slovakia. The result of this study was that seven national parks in Slovakia have potential for PAN Parks. The selected national parks are: Malá Fatra NP, Muránska Planina NP, Nízke Tatry NP, Tatra NP, Slovenský Kras NP, Poloniny NP, and Slovenský Raj NP. (VANČURA 2002).

Results of above mentioned study for Slovak republic are listed in the Tab. 2

Name of Country: Slovakia	
Number of areas researched	39
Total size of areas researched (ha)	1,043,778
Number of potential PAN Parks	7
Total size of potential PAN Parks (ha)	247,322
Total size of potential PAN Parks (%)	2.7

Tab. 2: Results of the PAN Parks Foundation study for Slovakia

One of the most obvious candidates for PAN Parks certification was Tatra National Park. The administration of TANAP had already had an ambition to meet the PAN Parks standards and at that time a Letter of Intent between the administration of TANAP and the PAN Parks Foundation was signed. Unfortunately the consequences of the windstorm in November 2004 were an obstacle in meeting the PAN Parks criteria. However, further development of TANAP was quite different and the approach to management of the national park prevented to continue in process of certification for PAN Park.

Slovenský Raj NP applied for PAN Park certification and as a candidate park cooperated for several years with PAN Parks Foundation to fulfill criteria for the certification. The park administration worked in that time on improving park management and visitor management. Slovenský Raj NP belonged to the first seven national parks that PAN Park Foundation chose from 20 candidates and they obtained the status of a candidate parks. One of the greatest problems of certification was the insufficient size of the wilderness. One of the main tasks of this process was obtaining the appropriate data about distribution, frequency, expectations and knowledge about services in order to develop the Visitors

Management Plan and Sustainable Tourism Development Strategy (LACKOVÁ 2007).

The main aim of this thesis was to asses the potential of the Tatra National Park for PAN Park certification, to find out the lacks and recommend further steps for increasing the national park's possibilities for PAN Park certification.

2.2 Methodology

The main approach used to find out the feasibility of Tatra National Park for PAN Park certification was assessment of its potential against PAN Parks Principles and Criteria. For the first feasibility assessment Principles 1 to 3 were chosen. These principles are called management/process principles and indicate the appropriateness of the park in applying for PAN Parks certification. The fulfillment of these Criteria represents the first step in PAN Park certification.

In order to collect the data for a proper assessment of PAN Parks potential the following methods were used:

- a survey of all available data and information about TANAP based on literature data, internet data and data collected from representatives of management bodies and authorities
- meetings and discussions with representatives of responsible bodies
- SWOT analysis

Data about the TANAP were collected basically from available literature and data: studies, papers, documents (internal or official), legislation, personal discussions, email communication, media releases and others.

SWOT analysis was used to reveal strengths, weaknesses, opportunities, and threats of the national park in connection with possible PAN Park certification. It is a relatively simple, but effective method to look at the present situation and future possibilities (Beunders 2002). Special emphasis in this analysis was given to the potentials of the area. SWOT analysis is a structured way to evaluate all aspects that are important in regards to the objectives. A difference can be made between internal and external aspects:

- internal – the park's strengths and weaknesses. These are items that can be found within the park boundaries or in the park organization.
- external – the opportunities and threats facing the park. Opportunities and threats are aspects that influence the situation of the park and can be both positive and negative.

2.3 Objectives and purposes

The main objectives of the present study were:

- to analyze the PAN Park potential and the possibilities of Tatra National Park in Slovakia;
- to asses feasibility of the park for PAN Park certification using assessment of natural values and management of the park;
- to recommend the steps and approaches that are needed for improvement, better effectiveness of the national park management and governance to apply for PAN Park certification.

3 RESULTS

3.1 General overview of Tatra National Park

The Tatra National Park (TANAP) is situated in the northern part of Slovakia. It is bordered by geographical co-ordinates of 49°05' - 49°20' northern latitude and 19°35' – 20°25' eastern longitude. The main part of TANAP is the Tatra Mountains, which are about 60 kilometers long and 17 kilometers wide. They include two geomorphologically different ridges – the West Tatra and East Tatra Mountains. The High Tatra Mountains, which are part of the Eastern Tatra Mountains and situated in the center of the national park are the highest mountains of Slovakia and the entire 1,500 kilometers of the Carpathian arc (VOLOŠČUK 1999).

The geological structure of the region reflects a complicated geological development in the past. There are three main structural units: the crystalline core, the sedimentary mantle of surrounding ranges, and sedimentary deposits of adjacent basins. The crystalline core forms most of the main ridge and the southern slopes in the granite West and High Tatra Mountains. This part consists mostly of metamorphic and granitic rocks.

The Tatra Mountains are significant also for their great relief, with an altitudinal difference of 800 to 1,800 m from the surrounding basins to the summits, of which 17 are higher than 2,500 m. The range is also of enormous importance for understanding the geology of the core mountains of the Carpathians, and provides excellent opportunities for studying glacial and periglacial geomorphological phenomena in alpine environments with an immense number of rocky peaks, deep valleys, and glacial lakes. In the context of the Carpathians, the Tatra Mountains might be compared partly (in geology and geomorphology) with the top-most ranges of the Rodna and Transylvanian Alps (VOLOŠČUK 1999).

The wide variety of soils in TANAP reflects numerous soil-forming factors, especially geological and geomorphologic conditions. There are several different types of soil within the area: immature soils (lithosols), rankers, rendzinas, cambisols (brown soils), pseudogleyed soils, and podzols.

The Tatra Mountains are a very important European watershed and waters from them drain in two directions: through the Dunajec and Vistula rivers to the Baltic Sea and through the Váh and Danube into the Black Sea. The most important mountain stream is Belá, which drains six valleys in the central part of the Tatra mountains. One of the specifics of TANAP is a very high density of glacial lakes. Totally there are 110 glacial lakes and three of them larger than 10 ha. Some of them are 25 m deep or deeper (MIDRIAK 2004). Sources of underground water have a very high importance as well and the area as a whole has a very important function in water retention.

The climate of the mountains is determined by their geographical situation and is characterized by a quite high degree of continentality. There are large temperature differences between summer and winter and there is high precipitation during summer months. The mean annual temperature ranges from 5.5°C at the lowest elevations to -3.8°C on the summits. The warmest month is July. Snow cover can be as much as 3 m high at the highest altitudes.

Vegetation structure is determined particularly by climatic conditions and six major altitudinal vegetation zones can be recognized (submontane, montane, supramontane, subalpine, alpine and subnival zones). Species richness of plant is very high. Overall 1,400 species of vascular plants were found in the area. Many of them are rare and threatened species and several of them are endemic for the Tatras (39 species), some of them for Western Carpathians (42 species) and some of them belong to Carpathian endemic species (57 species). Some of them are glacial relicts which have the southern border of their range there.

Fauna of TANAP is rich for boreal species and some of them are endemic. Overall, occurrences of 11 species of fishes and 2 species of Cyclostomata, 6 species of amphibians, 5 species of reptiles, 102 species of breeding birds and 42 species of mammals were found. Tatra ecosystems are rich for invertebrates as well and many of them belong to rare species with occurrence only within this area.

3.2 Natural values

The Tatra National Park was established on 1st of January 1949 by the Act of the Slovak National Parliament n. 11/1948 as the oldest national park in Slovakia. The Western Tatras were affiliated to the TANAP in 1987 by a Decree of the Government of the Slovak Socialistic Republic n. 12/1987. Specification of current borders and buffer zones was adjusted by the Decree of the Government of the Slovak Republic n. 58/2003 in 2003. In 2004, with the entrance of Slovakia to the European Union, TANAP was designated as a part of the NATURA 2000 network.

TANAP is considered to be a protected area of IUCN category II and it is inscribed in the UN List of protected areas. The National park was included into the UNESCO's Man and Biosphere Programme and was recognized as a Biosphere reserve together with Tatrzanski Park Narodowy in Poland in 1993. The area has also been identified as the core area of European importance of the Pan-European Ecological Network under the Council of Europe's Pan-European Biological and Landscape Diversity Strategy.

In total TANAP in its current borders covers an area of 73,800 ha. The buffer zone of the national park covers 30,703 ha. It is situated within two regions: Žilinský and Prešovský regions and three districts: Tvrdošín, Liptovský Mikuláš and Poprad. The boundaries of the national park were drawn widely enough in order to contain all ecosystems of the Tatra mountains. The borders of the Biosphere reserve cover the national park area with a core zone consisting of 49,444 ha, the transition zone (23,641 ha) and outer zone (32,575 ha).

TANAP covers all important habitat types and ecological processes, representing existing samples of natural ecosystems in the region. The area embraces several important and unique habitats with occurrences of several endangered, threatened and endemic species. Flagship species of TANAP are two endemic subspecies chamois (*Rupicapra rupicapra tatrica*), marmot (*Marmota marmota latirostris*) and golden eagle (*Aquila chrysaetos*). There are other important species such as large carnivores that occur very frequently in the area, have stable populations in this territory and are considered as a symbol of European wilderness, such as the brown bear (*Ursus arctos*), wolf (*Canis lupus*) and lynx (*Lynx lynx*).

Fig. 3: Tatra chamois (*Rupicapra rupicapra tatrica*) – flagship species of TANAP (copyright of all photographs http://www.arollafilm.com/)

NATURA 2000

The area was identified as a Site of community interests and included into the List of proposed SCI's of the Slovak Republic with code SKUEV0307 Tatry. This area covers 62,272.12 ha. The great number of natural habitats of community interest is to be found here and the conservation of them presents the main objectives of this SCI. These habitats are the following:

3130 Oligotrophic to mesotrophic standing waters with vegetation of *Littoreletea uniflorae* and/or *Isoëto-Nanojuncetea*

3220 Alpine rivers and herbaceous vegetation along their banks

6430 Hydrophilous tall-herb fringe communities of *Petasites*

4060 Alpine and boreal heaths

4070 Bushes with *Pinus mugo*

4080 Sub-Arctic *Salix* spp. scrub

6150 Siliceous alpine and boreal grasslands

6170 Alpine and subalpine calcareous grasslands

6230 Species-rich *Nardus* grasslands on siliceous substrates in mountain areas

6430 Hydrophilous tall-herb fringe communities of plains and of the montane to alpine levels

6520 Mountain hay meadows

7110 Active raised bogs

7120 Degraded raised bogs still capable of natural regeneration

7140 Transition mires and quaking bogs

7230 Alkaline fens

8110 Siliceous scree of the montane to snow levels

8120 Calcareous and calshist scree of montane to alpine levels

8160 Medio-European calcareous scree of hill and montane levels

8210 Calcareous rocky slopes with chasmophytic vegetation

8220 Siliceous rocky slopes with chasmophytic vegetation

8310 Caves not open to public

9110 *Luzulo-Fagetum* beech forests

9130 *Asperulo-Fagetum* beech forests

9140 Medio-European subalpine beech forests with *Acer* and *Rumex arifolius*

9150 Medio-European limestone beech forests of *Cephalantero-Fagion*

9180 *Tilio-Acerion* forests of slopes, screes and ravines

91E0 Aluvial forests with *Alnus glutinosa* and *Fraxinus excelsior*

91D0 Bog woodland

91Q0 Western Carpathian calcicolous *Pinus sylvestris* forests

9410 Acidophilous *Picea abies* forest of the montane to alpine levels

9420 Alpine *Larix decidua* and *Pinus cembra* forests

List of the species of European importance listed in Annex II. of the Council Directive 92/43/1992 (habitat directive) is following:

Plant species:

Pulsatilla slavica, Cypripedium calceolus, Campanula serrata, Tozzia carpathica, Cochlearia tatrae, Dianthus nitidus, Scapania massalongi, Mannia triandra, Tortella rigens

Fig. 4: One of the representatives of large carnivores – brown bear (*Ursus arctos*)

Animal species:

Carabus variolosus, Lampetra planeri, Triturus cristatus, Triturus montandoni, Bombina variegata, Rupicapra rupicapra tatrica, Lynx lynx, Lutra lutra, Ursus arctos, Myotis bechsteini, Canis lupus, Microtus tatricus, Marmota marmota latirostris, Barbastella barbastellus, Rhinolophus hipposideros.

SPA Tatry has been established by the Decree of Ministry of Environment of Slovak Republic n. 4/2011 and covers an area of 54,611.29 ha. Bird species listed in Annex I. of the Council Directive 79/409/EEC (bird directive) on the conservation of wild birds are as follows: *Aquila chrysaetos, Tetrao urogallus, Aegolius funereus, Tetrao tetrix, Glaucidium passerinum, Tetrastes bonasia, Falco peregrinus, Aquila pomarina, Ciconia nigra, Caprimulgus europeaus, Dryocopus martius, Picoides tridactylus, Lanius collurio.*

Fig. 5: One of the target species of SPA Tatry Eurasian Pygmy Owl (*Glaucidium passerinum*)

3.3 Habitat management

Management objectives and design of protected area according to law

The key document, which states the primary objective of the protected area, management and restrictions, is the current legislation and namely Act nr. 543/2002 on Nature and Landscape Conservation. The Act in its section 19 (1) defines a national park as an "*an area usually more than 1,000 ha, predominantly with ecosystems substantially unaffected by human activities, or with unique and natural landscape structures that form biocentres and the most significant natural heritage in which the nature conservation is of higher priority than other activities*".

The Act defines five levels of nature protection within the whole territory of Slovakia and specifically for all protected areas. Design of TANAP is based on this system of nature conservation. The Act further defines all restrictions and activities which are allowed and for which there can be exception from law accorded within different levels. The higher level of protection there is the more restrictions it contains. The fifth level is the highest and corresponds with wilderness area.

Roles and duties of administration

The administration of TANAP is done by an official body responsible for the management of NP. The administration is a part of national agency State Nature Conservancy of Slovak Republic, which is settled in Banská Bystrica and is an official scientific agency of the Ministry of Environment of the Slovak Republic. Administration of TANAP has three workplaces with the main office settled in Tatranská Štrba and three different departments (Department of Landscape

Management, Department of Ranger Patrol and Department of Conservation). The target activities of administration of TANAP are as follows:

- nature and landscape conservation

- ecosystem management

- realization of activities within Action plans for threatened species and small scaled protected areas

- formulation of scientific opinions for regional and national authorities

- supervising of legislative compliances

- research and monitoring of ecosystems and assessment of human activities impact to ecosystems

- environmental education

The TANAP administration is supervised by the State Nature Conservancy of the Slovak Republic and has no formal decision-making responsibility. These responsibilities are held by local and regional offices, self-governing regions and districts authorities, which usually use advice and scientific opinions made by the administration. Regional offices have general and specific statutory competencies set out in the Act about nature and landscape conservation.

Another important body in TANAP is the State Forest of TANAP, which belongs to the Ministry of Agriculture and continues to serve the management of the NP in several aspects at the operational level. It is a contributory organization which manages forests, buildings and facilities owned by the state within TANAP.

Earlier, until the middle of the 1990's both organizations formed a single authority body for administration of public responsibilities of TANAP that was supervised by the Ministry of Agriculture and Forestry. This single body was divided in 1996 into the above mentioned independent organizations, each supervised by different Ministry.

Fig. 6: Black woodpecker (*Dryocopus martius*) benefits from dead trees

Management plan

In the current situation there is no valid management plan for the area. The last valid management plan was prepared in the beginning of 1990's and approved by the government in 1991. However, this management plan was valid until 2000. Since that time there has been no legitimate management plan for the entire area. Since the 1990's there have been several important changes, including both administrative structure changes and legislative changes, which mean that there is from the beginning of 2000 great need for a new comprehensive management plan. During the last ten years the administration of TANAP was working on a new document and one the first drafts was developed in the 2004 (VOLOŠČUK ET AL. 2004). However this document wasn't finished and officially approved.

The management planning has a legislative background. According to paragraph 54 (point 5) of the Act 543/2002 the management plans form a basis for ensuring permanent care of protected areas and their protective zones. An approved management plans for a protected area or for an area of international interest are binding document for the development and approval of further management if they relate to the same area. These documents ensure sufficient permanent care of such an area. The management plan for the area of international interest always takes precedence over other management plans. Proposed measures of practical care are discussed by a body procuring the management plan with known owners of affected lands. According to paragraph 54 (point 17) of the above mentioned, law management plans of national parks and areas of NATURA 2000 network are developed by the Ministry and approved by the Government.

The content of the management plan is stated by the Decree of Ministry of Environment n. 24/2003 through which the law is carried. One of the problems of management planning in Slovakia is that there is still not a valid and officially approved methodology of working out management plans for protected areas. Thus the process of management planning is very slow and only a few protected areas in Slovakia have a valid management plan. Unfortunately this does not include any of the national parks or protected landscape areas.

Proposals for management plans according the Decree 24/2003 have stated long term and short term objectives. Proposals for management plans from the year 2004 determine long term and short term objectives for each zone of the national park (VOLOŠČUK ET AL. 2004).

Long term objectives for zone A are:

-   to maintain autoregulation, autoreproduction and autoregulation of all natural habitats and sustain their function and ecological stability;
-   to preserve the biodiversity of endangered, vulnerable, rare and endemic species in situ;
-   to maintain the aesthetics of unspoiled landscape without any not authentic elements.

Long term objectives for zone B are:

-   to strengthen the favorable status of habitat function that was partly disrupted by direct or indirect natural or anthropogenic factors;
-   to maintain biodiversity by supporting natural habitats, strengthening the ecological functions of ecosystems and setting up conditions for maintenance of the favorable status of ecosystems and endangered, vulnerable, rare and endemic species.

Long term objectives for zone C are:

-   to secure sustainable use of species and ecosystems by use of nature-friendly ways of management;
-   the improvement of the ecological stability of forest stands where species composition was changed by previous forestry management;
-   to maintain the favorable status of habitat with different functions such as medical and rehabilitation, scientific, educational, recreational and sport.

A rather important milestone in the last decade of TANAP was a great wind storm in November of 2004. This kind of event apparently occurs irregularly but frequently and is a natural part of natural disturbances within this region. During this event quite a big area of the forests was affected and fell down. The size of the area which was impacted extends more than 12,000 ha at altitudes between 700 m to 1,350 m which covered about 1/6 of the national park's area. The majority of fallen trees were situated outside of small scaled protected areas with the fifth level of protection. However some of the affected forests were included in national nature reserves or nature reserves as well. Quite a huge discussion about future of TANAP was raised at the time and the consequence of the next development was the removal of a majority of the fallen trees from the affected area. One of the recommendations of the IUCN mission in TANAP (CROFT ET AL. 2005) was that in the proposed zonation in areas with the fourth and fifth levels of protection no fallen or broken timber should be removed, and no artificial rehabilitation measures taken. In level 3 areas 50 % of the trees should remain, a buffer zone be created where all fallen and broken timber is removed, and some assistance to achieving natural rehabilitation can take place. However, the majority of the territory was managed by the approaches of forestry and even in the areas with forth of fifth levels of protection fallen timber was removed. Overall only 5 or 6 % of the whole affected territory was not cleaned of fallen trees.

A rather similar situation occurred at the end of 20th century in the Bavarian Forest National Park in Germany. Forests in the national park were hit by a smaller storms and followed by bark beetle gradation. Such events created in the national park a unique opportunity to investigate natural vegetation dynamics with little human interaction (FISCHER ET AL. 2002). Decisions made by the management authorities were to not clear away storm damaged trees and allow the forest to develop naturally. Due to these decisions one of the very important wilderness areas in Central Europe was established. This approach could be a very good example also for the Tatra National Park.

One of the biggest events during the discussions about future measures in TANAP was a public movement for the conservation of the national nature reserves, Tichá and Kôprová Valleys. Act 543/2002 enables legal exceptions from restrictions in protected areas. In these two valleys sanitary forestry actions were planned in 2007 to prevent bark beetle infection after the large wind storm and some work had been already started in that time. A large public protest organized by the NGO Wolf – Forest Protection Movement with the support of several other environmental NGO's was one of the most important protests in the history of TANAP. In this case the decisions of managers and local foresters were successfully influenced and these two valleys remained undisturbed. This protest proved the importance of the involvement and support of NGO's and the wider public in nature and wilderness protection.

Fig. 7: Wilderness zone of TANAP

The Tichá and Kôprova Valleys have become symbols of wilderness in Slovakia right after this event. Nowadays this area is home to more than 40 brown bears, one pack of wolves and many other rare and endangered species occuring in primeval forests and several natural habitats (BALÁŽ 2010).

Zoning and wilderness area

According to paragraph 30 (point 1) of the Act 543/2002, protected areas may be divided into a maximum of four zones based on the status of natural habitats if care of the protected areas requires it. The zones are determined and split into levels based on the character of their natural values so that the fifth level of protection applies in A zone, the fourth level of protection applies in B zone, the third level of protection applies in C zone, and the second level of protection applies in D zone.

In the current situation there is no zoning according to the above mentioned legislation and overall zoning of the NP is in a process of preparation. However, the current status is that the area is divided into different levels of protection and in fact this could be applied as the zoning of the area. This status is outdated and the need for proper zoning based on the current situation in the national park is highly urgent.

In the last years several proposals of zoning have been made by the administration of TANAP but none of them was officially approved. These proposals covered the whole territory of the NP. The last proposal from the year 2010 was made by the Ministry of Environment and aroused a great public movement against this proposal of zoning because it highly curtailed zone A and enabled huge investment activities and forestry in nature reserves with, up to this moment, the fifth level of protection. This proposal decreased the range of zone A to 8,000 ha and the proposal was not negotiated enough with all stakeholders, not even with the administration of the national park.

Fig. 8: Kôprová valley

In its current stage the wilderness area of the national park with the fifth level of protection covers 27 national nature reserves, 27 nature reserves, which represents more than a half of the territory of TANAP (38,551 ha – 52 %). Small scaled protected areas in the buffer zone with the fifth level of protection cover 128.6 ha. This network of small scaled protected areas within the NP constitutes an unfragmented complex of most important and valuable habitats and creates a very important base for building up a sufficient wilderness area. Since zoning is still in process the wilderness area ranges from 54 to 57 % according to different proposals.

Conservation of threatened and endemic species
Management of TANAP gives special attention to threatened, endangered and endemic species. Action plans for several of them were elaborated and management of these species followed these plans. These Action plans were elaborated for the chamois, marmot and golden eagle. All Action plans for these species are already completed.

The chamois belongs to the flagship species of TANAP and long-term data about population dynamics of the species are available. The population is regularly monitored by periodic counting, using telemetry as well. The most vital population of the species is in Belianske Tatry, which represents the most eastern part of TANAP.

Another flagship species are marmots. The population is also periodically long-term monitored. Since a quite rapid decrease of the local population in Belianske Tatry has been observed in the last years the administration of TANAP proceeded with a re-introduction project of this species into that part of the national park and this project was successful.

3.4 Visitor management

Tatra National Park belongs to the most visited protected areas in Slovakia. Historically, the area belonged to the favorite tourism destination in Slovakia not only for Slovak people but quite frequently for foreign tourists. The whole region was historically developed with the increased numbers of tourists and visitors to the Tatra Mountains with increasing interests in different sport and recreational activities in these mountains.

Numbers of visitors

The number of TANAP visitors has been regularly monitored since 1972 and it was continuously increasing. These data are obtained during one day in the summer season, usually at the beginning of August when the attendance of the Park is at its highest. Overall, there are 57 check points within the whole Park and visitors are counted within 98 hiking trails. Data from the last decade are shown in Fig. 3.

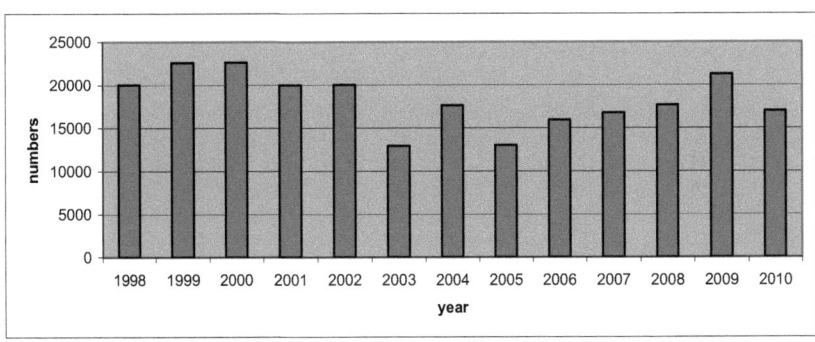

Fig. 9: Numbers of visitors of TANAP during the last decade

Generally the number of visitors can reach more than 2 million annually. However, the highest numbers of visitors occured in the period of 1982 – 1987 when annual attendance reached 5.5 million visitors and in the period 1990 – 1993 it dropped down to 3.5 million visitors annually (Švajda 2002). The highest pressure is usually in the central part of the High Tatras where the town of Vysoké Tatry is situated, but also in the settlements of Štrbské Pleso, Starý Smokovec and Tatranská Lomnica, which are the most frequently visited places in TANAP. Less visitors usually visit the western part of TANAP and in the eastern part – Belianske Tatry, which was for some years closed to tourists; currently there is only one hiking trail, which is not free of charge access (Monková valley).

Carrying capacity

There are several studies dealing with the carrying capacity of TANAP. Some of them have stated the carrying capacity of visitors (MIDRIAK 1989). The study enumerate the real carrying capacity of alpine zone with regards to nature and soil conservation and geology at 11,645 to 14,240 visitors per one day, which means annually from 1 to 1.4 million visitors per summer season. Some studies, which are focused only on certain regions of TANAP, have more detailed analyses per existing trails. For instance the hiking trails in Belianske Tatry can be classified into three categories: very high carrying capacity with proposed attendance of 500 persons/day at maximum, middle carrying capacity with proposed carrying capacity of 300 persons/day and low carrying capacity with maximum attendance of 100 persons/day (BARANČOK AND BARANČOKOVÁ 2008).

Visitor management plan and Rules for visitors

Currently, there is no independent visitor management plan in the sense of principle 3 of the PAN Park Principles and Criteria. Some principles and rules of visitor management were implemented in proposals of the overall management plan of TANAP. Rules for visitors represent a valid document for regulation of visitors and tourism activities. This document was prepared at the end of the 1990's and has been valid since 1999. The document was approved by the Regional office for environment of Prešovský Region as an official Decree n. 1/1999 which adjusts details and specifies duties of national park visitors, time and spatial scales of different activities that are allowed in different parts of the protected area. However, it should be pointed out that nowadays it doesn't fulfill the needs for the management of visitors.

The Rules for visitors specify possibilities of tourist, recreational and sport activities within the area of the NP. Generally, there is a restriction of entrance to the NP and visitors can enter the park along marked hiking trails. Several of them have seasonal prohibition of entrance from 1st of November to 15th of June. This regulation concerns 37 marked hiking trails. Furthermore, the document specifies localities and places reserved for skiing, ski alpinism, cross country skiing, climbing, alpinism and mountaineering, sailplaning, paragliding and conditions for all these activities. Entrance to the Park out of marked trails is possible only with a certificated guide excluding seven specific sites, which are described in the document. There is also a limitation: only 30 persons can reach some of the summit that is not included in marked trails. One certificated guide can guide only up to five people at the moment. Climbing and skialpinism is permitted for all members of climbing or skialpinism associations excluding eight specified localities for climbing and six specified localities for skialpinism. All these limits represent some kind of visitor zoning system that allows activities and time periods for each zone of the NP.

Services for visitors

TANAP offers overall 600 km of hiking trails on 69 marked trails, 16 marked and maintained bike trails and four spots reserved for paragliding. The other available sport activities within the NP are skiing and cross country skiing, climbing and mountaineering, skialpinism, rafting and horse riding

The administration of TANAP offers some services provided by administration staff, rangers and scientific experts of PA. These services include guided tours along marked trails with a professional ranger or specialist with expert commentary on to nature conservation or specific fields like geology, vegetation and botany or zoology. There are also possibilities for guided tours into wilderness outside the marked trails to observe rare birds or large carnivores. Other possibilities are rafting on the Belá River or horse riding outside the NP, including visits to some natural reserves in the vicinity of the NP. All these activities are also one of the additional sources for the financing of the NP.

Another possibility for visitors is guided tours into alpine habitats outside the marked trails with a special trained guide. These guides are able to guide anywhere in the alpine zone and usually they are needed for hiking into some of the summits including the highest peak of Gerlach.

The administration of TANAP doesn't have a visitor centre, but some information centers are administrated by State Forests of TANAP. This organization owns a Museum of Tatra Nature in Tatranská Lomnica, which is a visitor centre and a Botanical Garden situated in the same settlement. There are six educational trails in the territory of the NP that contribute to the information and education of visitors.

activity	climbing	Ski alpinism	camp places	ski centres	cross country skiing **	biking trails **	hiking **
numbers	whole area*	6	1	7	108	150	690

Tab. 3: Number of localities for different sport activities
  * - excluding 8 localities where climbing is prohibited according the Rules for visitors
  ** - in these cases of cross-country skiing, biking and hiking numbers mean total length of marked trails for particular activity

activity	wilderness area	other zones of NP
hiking	only on legally designated path or with the guide	on legally designated path in level 4 and 3
snow shoe hiking	only on legally designated path or with the guide	on legally designated path in level 4 and 3
cross-country skiing	only on legally designated path	on legally designated path in level 4 and 3

climbing	allowed on the whole territory excluding 8 sites where it is prohibited by Rules for visitors	
ski alpinism	allowed on specified 6 localities	
biking	only on legally designated trails	on legally designated path in level 4 and 3
bird watching	only on legally designated path or with the guide	on legally designated path in level 4 and 3
wild life watching	only on legally designated path or with the guide	on legally designated path in level 4 and 3

Tab. 4: Sport and recreational activities allowed in wilderness zone and other zones of TANAP

3.5 Assessment of Tatra National Park potential

Assessment of TANAP against selected Principles and Criteria of PAN Parks is analyzed in the following table:

Principle 1: Natural values			
Criterion	**Indicator**		**Fulfillment**
1.1	The area protected by means of an act or decree		+
1.2	International recognition		
	1.2.1	IUCN classification, Biosphere reserve, etc.	+
	1.2.2	NATURA 2000 sites	+
	1.2.3	Important habitat and species	+
1.3	The minimum size of the PA is 20,000 ha		+
Principle 2: Habitat management			
2.1	Design of the PA aims to maintain natural ecological values		
	2.1.1	Maintenance of natural ecological values as a management objective	+

	2.1.2	The design of the PA allows all key natural values to exist and be maintained	+
	2.1.3	Evidence of bio-geographical connections within the PA, with its adjacent areas, and with other PA's	+
2.2		Regulations protecting the area are adequately enforced	+
	2.2.1	The protected area management has clearly defined roles/duties and can halt activities threatening the conservation values within the PA	+ -
	2.2.2	The protected area management has regulations clearly defining the roles/duties of all relevant agencies	+
2.3		The PA has a long-term conservation strategy that is actively implemented through a management plan	Not finished
	2.3.1	Conservation strategy implemented through subplans	+ -
	2.3.2	The conservation strategy/management plan developed through a planning process that includes procedures for revision and approval and the participation	+ -
	2.3.3	Links between nature conservation management, the visitor management and regional STDS	+ -
	2.3.4	The conservation strategy/management plan has long- and short –term goals	+
	2.3.5	The conservation strategy/management plan goal is maintaining of ecological processes and biodiversity	+
	2.3.6	The conservation strategy/management plan includes research programmes	+
	2.3.7	The conservation strategy/management plan includes programmes for improving the socio-cultural and economic benefits of the PA for surrounding communities	+ -
	2.3.8	The conservation strategy/management plan is based on an adequate site assessment	+
	2.3.9	The conservation strategy/management plan is addressing needed capacities to effectively managed	+ -

		the PA	
	2.3.10	The conservation strategy/management plan is addressing existing and future external and internal threats and pressures to the PA	+
	2.3.11	The conservation strategy/management plan is successfully implemented	+ -
	2.3.12	Monitoring of management effectiveness	+
2.4		PA management make use of zoning or other system that achieves conservation strategy	Not finished
	2.4.1	Zoning system that ensures effective protection of the PA	+ -
	2.4.2	The zoning is based on a clear method of demarcating boundaries, both around the PA and in between its zones	+
	2.4.3	The zoning system allows human activities compatible with the conservation strategy	+
2.5		The PA has an ecologically unfragmented wilderness area of at least 10,000 ha where no extractive uses are permitted	Not finished
	2.5.1	Ecologically non-fragmented wilderness area of at least 10,000 ha	+
	2.5.2	The management plan includes a clear management strategy and plan for managing the wilderness area at long term	+ -
	2.5.3	Ecological processes within the wilderness area are undisturbed, those missing are under restoration	-
	2.5.4	The plan includes a description of all human activities permitted within the wilderness area (goal, scale, impacts)	+ -
	2.5.5	All human activities permitted outside the wilderness area neither conflict with nature conservation goals of wilderness area	+ -
2.6		If the PA is not zoned, management of the whole area aims to maintain key natural ecological processes	
2.7		Activities in the area surrounding both the PA and the wilderness	

		area don't adversely impact on the conservation objectives of these areas	
	2.7.1	An adequate buffer zone inside or around the PA	+
	2.7.2	The adequate connectivity between the PA and undisturbed areas outside the PA	+
2.8		The PA management system pays particular attention to threatened and endemic species and habitats, and ecosystem dynamics	
	2.8.1	The management plan and other sources provide information on endemic, red-listed, vulnerable or other rare species	+
	2.8.2	Management measures are being taken to mitigate the main threats to the species mentioned above	+
	2.8.3	Introduction and re-introduction projects are implemented in the PA	+
	2.8.4	Knowledge on invasive alien species occurring in the PA, respective management measures have been taken	+
	2.8.5	Habitat or ecosystem restoration plan	+
2.9		The nature management plan includes training programmes for staff	
	2.9.1	The park has a training programme for the staff and others involved in nature management	+
	2.9.2	The programme is systematically monitored	+ -
2.10		Two or more PAs adjacent to one another within or across a national borders, co-operation in management is actively sought	
	2.10.1	The bio-geographical connections of the PA with adjacent (transboundary) PA's	+
	2.10.2	Cooperation with adjacent PA's	+
Principle 3: Visitor management			
3.1		The PA has a visitor management plan, which safeguards the natural values and is actively implemented	
	3.1.1	The visitor management plan, resulting from a	+ -

		planning process that included the participation of different parties and a procedure for revision and approval	
	3.1.2	The plan has long- and short-term objectives, is communicated to different target groups	-
	3.1.3	The visitor management strategy aims at measures to avoid or limit negative impacts of visitors on natural values and ecological processes	+ -
	3.1.4	A visitor zoning system, specifies visitor access, allowed activities and time periods for each zone	+
	3.1.5	The needs for and the effects of implementing plan are systematically monitored	+ -
3.2		Under the visitor management plan visitors are offered a range of high-quality services, facilities and activities	
	3.2.1	The visitor management plan leads to range of activities preferred, services to be provided, facilities to be installed	+ -
	3.2.2	The visitor information points have clear goals and a policy, different communications and promotion techniques, and are regularly open and accessible at appropriate point	+
	3.2.3	Visitors have specific opportunities to observe and experience wildlife and other natural features of the PA	+
	3.2.4	Specific monitoring surveys the number and type of visitors, their activities, their use of facilities and services, and their level of satisfaction	+ -
	3.2.5	Based on visitor satisfaction, the quality of the activities, services and facilities provided are steadily improved	+ -
	3.2.6	There are existing and planned partnerships with tourism providers, communities and other partners in use	+ -
	3.2.7	The safety regulations concerning activities and the use of facilities are enforced, monitored and updated	+ -

3.3		Visitor management creates understanding of and support for the conservation goals of the PA	
	3.3.1	There are different visitor target groups that need to understand and support the conservation goals of the PA addressed by specific messages and different techniques	+
	3.3.2	A code of conduct for visitors is communicated to all visitors. specifying for which visits a qualified guide is needed	+
	3.3.3	The PA has a communication and marketing plan that is successfully implemented in communication with the tourism marketing of the surrounding region	+ -
3.4		The visitor management plan includes training programmes for staff and others involved in the provision of visitor services	
	3.4.1	The training programme addresses the staff and others involved in offering activities, services, and facilities to visitors	-
	3.4.2	It specifies goals, target groups, methods, and timing of these trainings	-
	3.4.3	The training needs of staff and other people involved are assessed and the result regularly monitored	-

Tab. 5: Assessment of TANAP against selected Principles and Criteria of PAN Parks

Legend: +   indicator is fulfilled
      + - indicator partly fulfilled
      -   indicator not fulfilled

3.6 **SWOT analysis from wilderness management point of view**

Strengths
- high concentration of biodiversity and natural values on relatively small area;
- presence of many endemic species including chamois subspecies (*Rupicapra rupicapra tatrica*) and marmot (*Marmota marmota latirostris*) as animal flagship species and several plant species;
- relatively large proportion of possible wilderness area consisting of an unfragmented area compared to other national parks in Slovakia and the presence of large carnivores such as the bear, wolf and lynx;

- Tatra mountains as a national symbol of Slovak people with the highest peaks within whole Carpathian mountain region, and well known in foreign countries;
- very high potential for tourism related to high concentration of natural values in the area of the NP and historical and cultural values in the vicinity.

Weaknesses
- uncompleted process of the zoning of the area and unfinished valid management plan for the NP;
- the TANAP administration is not a formal decision-making authority;
- there are two official bodies responsible for the management of TANAP – administration of TANAP and State Forests of TANAP;
- lack of integrating approach to strategy and management of the NP, outranking of short term economic interest above environmental interests and nature conservation;
- negative impact of human activities in the past, high proportion of artificially forested areas with same age spruce stands which are endangered by wind storms and bark beetle gradation, this creates very contradictory opinions and approaches from different stakeholder groups.

Opportunities
- very high potential for the development of sustainable ecotourism mostly in the field of trekking, hiking, biking and other sport activities in the area of the National Park and adjacent region. Improvement of current situation with tourism and recreational activities towards sustainable ecotourism;
- very high potential for studying evolutionary and ecological processes of mountain ecosystems and possibilities for development of ecotourism based on the observations of restoration processes after wind storm calamity. High possibilities to improve awareness about importance of wilderness management and importance of an ecosystem-based management of the National Park;
- possibilities for improvement of quality of management of the NP connected with PAN Park certification. Increasing prestige, interests and willingness to visit area with PAN park certification;
- effective forms of communication (partnership, cooperation and participation) with key stakeholders on local and regional level in decision-making process and possibilities for establishment of good governance of protected area;

- solving of problems with ownership and financial compensation for landowners for restrictions resulting from legislation, enabling of closer cooperation with landowners and land users;
- strengthening the participation of local municipalities, landowners, key stakeholders and public in nature and landscape conservation.

Threats
- forestry management focusing on elimination of wind storm consequences especially in areas with highest level of nature conservation in nature reserves;
- high level of recreation, so called mass tourism and some sport activities and destruction of mountain ecosystems by development of big ski centers;
- disturbance of animals especially with some activities such as paragliding, skialpinism, snowboarding, field car or field motorcycles driving and snowmobile driving out of areas that are set aside for these activities. Disturbance of animals especially in critical phases of life cycle (winter time, reproduction period) could have very negative impacts (Huba 2005);
- huge construction development and development of infrastructure with high pressure upon mountain ecosystems. Such development is usually part of extensive investment and development projects, e.g. enlargement of ski resorts and ski slopes leads to permanent loss of biodiversity and important mountain habitats. Technical snowing of ski slopes leads to negative impact to water regime and diversity and landscape quality, e.g. decreasing of the water level in Štrbské Pleso lake.

4 DISCUSSION AND RECOMMENDATIONS

The Tatra National Park is one the national parks of Slovakia that has quite a high potential for PAN Parks certification. The protected area has a lot of natural values and high biodiversity concentrated in relatively small area. Another advantage is the relatively unfragmented and quite large size of area without interferences included in the fifth level of protection, which is a very good platform for wilderness area.

There are still some lacks in legislation regarding the position of national parks in Slovakia. IUCN guidelines clearly define the national park as category II. The current legislation defines the national park and in some parts this definition meets the criteria for the protected area, however it doesn't fulfill completely all objectives for this IUCN category. The IUCN mission in 2005 found that there are two critical points in the definition. One is the exploitation of natural sources where the legislation doesn't define the area in which the primary objective should be secured. The Act about nature conservation enables some interventions in the highest level of protection. The second critical point was a provision of environmentally and culturally compatible spiritual, scientific, educational, recreational and visitor management (CROFT ET AL. 2005).

All legislative issues are in their current status included in the Act about nature conservation. In order to strengthen the protection status of TANAP it would be suitable to establish an independent Act for the national park which would define all objectives, importance and regulations within the national park.

Currently the new law about nature conservation is in a process of preparation and it is likely to change many things. One of the important changes will probably be a new system of grades and zoning that would more suit IUCN requirements.

Habitat management
There are several problems in TANAP which have to be solved to improve management of the National Park and increase its possibilities for PAN Park certification. Many of them are regarding the second principle of PAN Parks and deal with criteria concerned with habitat management.

One of the biggest problems of management of TANAP is the separation of its management into two different organizations where each of them belong to a different ministry. Thus an integrated approach to management of TANAP is missing and moreover there are several administrative levels each with different competencies.

There were several suggestions about improving the administrative arrangements and the structure of decision making with respect to TANAP. However, generally it would be fine and useful to also concern other national parks in Slovakia. Recommendations of the IUCN mission in 2005 suggested that the best solution would be to reflect the best international experience and establish a more integrated management body with a non-executive decision-making board representative of all interests and a varying level of delegation of responsibilities from higher state authorities and covering state, charitable and privately owned land.

The best option for management of TANAP seems to be one official body responsible for all aspects of the NP, including nature and landscape protection, forestry and land use. Within this scenario the present administration of TANAP and the State Forests of TANAP would be fused into one management body, which should be established with executive responsibility for all matters of the National Park. An important part of this process should also be the establishment of two boards: a scientific and an advisory board. The advisory board would represent all of the interests in the area: state, regional and local government, local communities, land owners and land users and environmental non-governmental organizations active in the area. This model would be a very good example of good governance for the National Park and also for PAN Parks certification and suitable for all national parks in Slovakia.

Another big deficiency for PAN Park certification is a lack of management plan and proper zoning. This process takes a long time and it is still unfinished even though it started several years ago. It should be stressed that this is not only an important issue for PAN Park certification but generally a management plan is a very important tool for management of the NP. A management plan is a product of a planning process, documenting the management approach, the decisions made, the basis for these, and the guidance for future management (THOMAS AND MIDDLETON 2003).

Finishing the management plan and zoning of the area is one of the most important and crucial issues which should be done in the TANAP. The conservation strategy should be part of this document, which will help further management of the NP. One important issue is defining primary objectives according IUCN principles for category II. This primary objective should occur in at least three-quarters of the area (DUDLEY 2008, CROTF ET AL. 2005). A recently approved management plan and zoning is the most important priority and challenge for the TANAP administration together with proper financing and personal capacity building (S. Celer, pers. Comm.).

Generally, according to recommendations of IUCN (CROTF ET AL. 2005) the zoning system in the TANAP should be based on the principles of the IUCN category system. It should be a tool for maintaining all values of the NP, for implementing the agreed management strategy and plan, and as a basis for action on the ground, including the degree of intervention. The current system is not appropriate and needs a great revision. There are still settlement parts of the NP and some parts of settlement are in the fifth level of protection, which doesn't guarantee proper protection of these parts in the meaning of the fifth level.

There is also a need for the management plan of the wilderness zone that can be a part of the overall management plan, but PAN Parks Foundation prefers it to be a separate document that defines all objectives and management of a wilderness area. Passive management is an important tool, but should be actively planned for and included into this document (PAN Parks 2008). The management plan should ensure a clear strategy for the wilderness area with no exceptions for cutting and removing of trees and any interventions in natural processes. This principle has been quite significantly broken several times in the last years when interventions following the wind storm in 2004 led to the removal of fallen and broken trees for sanitary logging in the areas with the fifth level of protection. This was also the case of the Tichá and Kôprová Valleys.

ŠVAJDA (2009) in his thesis evaluated integrated protected area management in national parks in Slovakia using IPAM toolbox. According to his results TANAP has several gaps in management planning. The results of the analysis showed that there is a need for the improvement of the implementation phase, the phase of detailed planning and more specifically the deficiencies in basic concepts, and the design of regional economic programs, specifically planning. However, there are also some gaps in ecosystem-based management planning, personnel and organizational development, the evaluation of management effectiveness, and the development of PA's region. Studies presenting the economic impact of the National Park on the adjacent region are still missing. Economic evaluation was done for the adjacent Tatrzanski Park Narodowy (GETZNER 2009). The study has highligted the recreational benefits of the park as the most important and thus the National Park bears an eminent importance to the national economy.

One of the insufficient fields in the study of ŠVAJDA (2009) was financial planning. Generally, financial planning is still lacking in Slovakia. National parks as a part of the State Nature Conservancy of the Slovak Republic are financed through this governmental agency and in the last years are facing lack of finances. Thus, the administrations of national parks are almost completely dependent on a state budget and the majority of this money comes to operational costs of administrations. Therefore and not only for this particular issue the independence of TANAP administration is very important.

One of the problems dealing with the financing of TANAP is the compensation of private land owners, namely private and municipality forest owners that claim for loss of income. This problem is quite urgent because private owners, not only in TANAP but generally in national parks in Slovakia, have been ignored on a long term basis. This situation is due to the fact that TANAP and several other national parks in Slovakia were established during the communist regime with a top-down approach. There was no, or only a very poor, discussion process with relevant stakeholders in that time. The reason of this approach is that local residents usually have quite a negative attitude to national parks. According to interviews with representatives of TANAP, the attitude of private forest owners is improving and this could lead to a successful acceptance of TANAP zoning.

Therefore one of the important issues of finance planning is the adoption and implementation of a system of compensation for the elimination of opportunities for income generation by landowners. One of the possibilities is a system of compensations. However, a better option seems to be a system of rental of private owners land. It requires just setting fixed prices for particular forest habitats. Another possibility is the repurchasing of private land by the state. These approaches need proper financial analysis of all possibilities and scenarios.

Visitor management

Visitor management is one of the crucial issues in the management of PAN Parks. TANAP is the one of the most important tourist destinations in Slovakia. Tourism and recreation are found to be the second most significant threat and pressure to the natural values of TANAP. This result was shown also in SWOT analyses, that some of the pressures coming from tourism can be very serious. Visitors are usually not equally distributed within the area of the NP. The main concentration of visitors is usually in the High Tatras and especially in tourist centers in or around settlements such as Štrbské Pleso, Starý Smokovec and Tatranská Lomnica. Such distribution and increased numbers of visitors creates quite significant pressure to this part of the NP. One of the main threats is a demand for the expansion of existing and development of new sport facilities and tourist infrastructures.

TANAP doesn't have any specific visitor management plan in the sense that is required by PAN Parks Foundation. Similarly, in the case of other subplans, some parts dealing with visitor management can be incorporated into a general management plan of the NP. However, for the reason of PAN Park certification it is necessary to develop an independent visitor management plan, which should result from a planning process, and that would include the participation of different parties and a procedure for revision and approval.

Rules for Visitors are currently the most important tool in the management of visitors in the NP. The document should be reviewed and updated in accordance to specific natural values and TANAP conservation requirements. This necessity for reviewing of Rules for Visitors was stressed also by representatives of administration of TANAP. In order to cope with the growing number of visitors and increased demand for particular sports and recreational activities, necessary tools to control and regulate the number and the behaviour of visitors has to be improved.

The high number of visitors could make it difficult to achieve the PAN Parks standards since these areas are usually under a stronger pressure from human influence. Study of TAYLOR (2004) showed that PAN Parks as a whole receive a relatively low number of visitors compared to other categories or labels of national parks. This fact could likely be part of the reason why these areas are able to achieve PAN Parks' standards. Another indicative fact could be that PAN Parks are usually situated in areas reasonably distant from areas of high human habitation.

Increasing recreational use of national parks and protected areas can impact natural and cultural resources and the quality of the visitor experience. Determining how much recreational use can ultimately be accommodated in a park or protected area is often addressed through the concept of carrying capacity (MANNING 2002). The problem with TANAP is so called mass tourism. The previous studies showed that carrying capacity is annually from 1 to 1.4

million visitors for the summer season (MIDRIAK 1989). According to that study, the number of visitors in TANAP highly exceeds its carrying capacity, especially in the environment of the alpine zone. There is still a lack of some mechanism and tools to regulate the numbers of visitors. Some of them are used in Rules for Visitors with some seasonal access restriction into some valleys and on some hiking trails. ŠVAJDA (2002) also suggested some other tools. One of them is to locate and offer sport and recreational activities in the foothills of the Tatras where the resilience of the landscape is higher and where there are not so sensitive habitats. Another possible mechanism is to set up an entrance fee, at least for some hiking trails, which could possibly regulate the numbers of visitors. This issue has been the subject of discussion for quite some time and as a solution is rarely used; there is only one hiking trail with paid entrance in TANAP, in the eastern part of park in Belianske Tatry. This tool could be not only useful for the regulation of attendance to the NP but would also help diversify the financing of the NP and could be one of the sources for income. There is a similar situation in neighboring Tatrzanski Pak Narodowy in Poland and this mechanism is used in many national parks worldwide.

It is necessary to ensure that sport and tourist activities in the case of TANAP should retain within existing areas a trail established for these activities and in accordance with Rules for Visitors. It is not acceptable to enlarge existing ski slopes, especially into areas with fourth or fifth levels of protection, nor into areas within NATURA 2000 sites or with habitats of European importance. However, this was done in the last few years. Some of the ski slopes were enlarged towards places of some important habitats.

Generally, demands for economic development should be carefully assessed and follow the principles of zoning of an area. The situation after the wind storm in 2004 has especially led to huge investments and infrastructure developments that have increased the pressure to mountain ecosystems.

PAN Park certification is quite a big challenge for the TANAP, however the tourism should meet quality standards and be based on sustainable principles instead of mass tourism. The requirements for PAN Park certification are similar to recommendations of IUCN mission from 2005 (CROFT ET AL. 2005) in that there should be no further tourist infrastructure in zones A and B that suits the 5th and 4th levels of protection. There should be a greater focus on improving the quality of existing infrastructure instead of building new one.

Another important issue in which TANAP has some gaps is the implementation of conservation goals in communication. This strategy should be included in communication and marketing plans, which enables better communication with the tourism marketing in the surrounding region. This is closely connected to further steps in PAN Parks certification according to another PAN Parks principles and leads to the elaboration of a sustainable tourism development strategy.

PAN Park certification is challenging also in terms of the participation of local municipalities, landowners, key stakeholders and the public in nature and landscape conservation. The TANAP administration should, in case of certification, create or improve existing partnerships with tourism providers, communities and other partners. One of the further steps in this process is the creation of local PAN Park groups and a sustainable tourism development strategy. According to a study BERG ET AL. (2004) development of sustainable tourism in the area, implementing the STDS, has an essential role in increasing the dedication and action of both authorities and local peoples. This will contribute to increasing the tourism potential of the region with respect to the environment.

5 **CONCLUSION**

This study has identified the potential of the Tatra National Park for PAN Park certification. The National Park has a lot of natural values and high biodiversity concentrated in a relatively small area and a relatively unfragmented and quite large size of area without interferences, which is a very good platform for a wilderness area and creates the potential to become a certified PAN Park.

There are several lacks and problems which should be solved to increase its possibilities to apply for PAN Park certification at the first step. One of the most important tasks in this phase is the finalization of a management plan and the process of zonation, which should be officially approved. The management plan should consist of and fulfill all the criteria and indicators of PAN Park principles. There should be links between nature conservation management, the visitor management and regional sustainable tourism development. Management planning should be improved in several ways but mostly in the field of participation and stakeholders' involvement. This way TANAP could become a model in effective management through modern planning and governance arrangements. Currently the level of communication and participation of stakeholders is still insufficient and misses some platform for involvement of them in the planning process. It could be improved by establishing a consultative or scientific board of the National Park. In further steps, creation of a local PAN Park group can lead to the improvement of participation as well.

Zoning was found to be one of the crucial issues that should be finished as a consensus of all relevant stakeholders and officially approved. It should be based on the principles of IUCN Category system and should be a tool for the maintenance of all values of the National Park. The most important task is designation of the wilderness zone without any interventions into habitats and with sufficient size to protect the natural systems and processes on which the status of individual species and their habitats is depend. In accordance with PAN Parks' principles the wilderness zone should be coherent and unfragmented and as large as possible in the current situation.

There are many strengths in the current situation of TANAP and opportunities which bring PAN Park certification to the National Park. On the other hand there are also some weaknesses and threats which could negatively influence the process of verification and the final result. However, the strengths and opportunities should overwhelm existing weaknesses which should be improved. Threats should be minimized or in the best case completely eliminated. It is hard to say if it is possible to eliminate all threats. There are several of them closely connected with high tourism development and it could be quite difficult to improve the situation and develop a sustainable way of tourism. The development of sustainable tourism seems to be the right option for Tatra National Park's future.

There is the question of whether PAN Park certification is the right option for Tatra National Park. From the view of nature and wilderness conservation it is the best option for the National Park. PAN Park certification brings to certified protected areas several benefits including improved wilderness conservation, demonstration that the protected area is managed with high standards of quality management, and effective expertise exchange of knowledge and international recognition. Other benefits are in the field of tourism. PAN Parks certification provides effective tools to develop sustainable tourism, and control and monitor tourism, which is very important task for many protected areas. Finally, one of the very important benefits is the improvement of cooperation with stakeholders and local communities, and through branding and promotion activities opening up access to new markets for small businesses.

Generally, the Tatra National Park has quite a big potential for PAN Park certification. However, many issues in management should be improved or finalized and officially approved. PAN Park certification is quite a big challenge for the National Park and is highly dependent on highly qualified decisions on different levels that would lead to changes in management and governance of the National Park.

6 REFERENCES

6.1 Literature

BALÁŽ, E. (2010): The last fortress. Fifteen years with bears. ADIN, s.r.o. pp. 237 (in Slovak).

BARANČOK, P. & M. BARANČOKOVÁ (2008): Evaluation of tourist path carrying capacity in the Belianske Tatry Mts. Ekológia (Bratislava) 27 (4): 401-420.

BERG, CH. VAN DER, F. VAN BREE & S. COTRELL (2004): PAN Parks implementation process: cross cultural comparison – Bieszczady & Slovenský Ráj National Parks. In: SIEVÄNEN, T., ERKKONEN, J., JOKIMÄKI, J., SAARINEN, J., TUULENTIE, S., VIRTANEN, E. (eds): Proceedings of the Second International Conference on Monitoring and Management of Visitor Flow in Recreational and Protected areas, June 16-20, 2004, Rovaniemi, Finnland, Finnish Forest Research Institute, Working Papers.

BEUNDERS, N. (2002): PAN Parks' manual for sustainable tourism and development. Practical guidelines for the formulation of a sustainable tourism development strategy (STDS) for protected areas and their regions. Internal document of the PAN Parks Foundation. Netherlands Institute for Tourism and Transport Studies, Breda, The Netherlands, 100 pp.

COLEMAN, A. & T. AYKROYD (eds.) (2009): Conference Proceedings: Wild Europe and Large natural habitat Areas, Prague 2009, Wild Europe, European Comision and EU 2009

CROFTS, R., ZUPANCIC – VICAR, M., T. MARGHESCU & Z. TEDERKO (2005): IUCN Mission to Tatra National Park, Republic of Slovakia, April 2005, 43 pp.

DUDLEY, N. (ed.) (2008): Guidelines for Applying Protected Area Management Categories. IUCN, Gland.

ENGELDORP, B. VAN (2002): PAN Parks Verification under Construction, PAN Parks Foundation.

FISCHER, A., LINDNER, M., C. ABS & P. LASCH (2002): Vegetation dynamics in Central European forest ecosystems (near-natural as well as managed) after storm events. Folia Geobotanica 37: 17-32.

GEZNER, M. (2009): Economic and cultural values related to Protected Areas. Part A: Valuation of Ecosystem Services in Tatra (PL) and Slovensky Raj (SK) national parks. Final report. WWF.

HUBA, M. (2005): Fauna of Tatra Mts. in the context of sustainable development of the Tatras region. Folia faunistica Slovaca, 10 (10): 45-53. (in Slovak)

JONES-WALTERS, L. & K. ČIVIĆ (2010): Wilderness and biodiversity. Journal for Nature Conservation 18: 338-339.

KUN, Z. (2002): Potential PAN parks in Europe: A quickscan of European protected areas against selected PAN Parks indicators. PAN Parks Foundation

LACKOVÁ, V. (2007): Example of the active participation on "Visitor management plan" – one of the principles to become PAN Park (case study in Slovenský Ráj National Park). Thesis. Technická Univerzita vo Zvolene. (in Slovak)

MANNING, R.E. (2002): How Much is Too Much? Carrying capacity of National Parks and Protected Areas. Pp. 306-313 In: ARNBERGER, A., C. BRANDENBURG & A. MUHAR (eds.): Monitoring and Management of Visitor Flows in Recreational and Protected Areas. Conference Proceedings.

MARTIN, V.G., KORMOS, C.F., ZUNINO, F., MEYER, T., U. DOERNER & T. AYKROYD (2008): Wilderness Momentum in Europe. International Journal of Wilderness. 14 (2): 34-43.

MIDRIAK, R. (1989): Load limits on tourist footpaths in the Tatra National Park with regard to the destruction of their surface. Zborník prác o Tatranskom národnom parku 29: 239-251. (in Slovak)

MIDRIAK, R. (2004): Mountain areas and their sustainable development. Technická univerzita vo Zvolene, 174 s. (in Slovak)

PAN Parks (2008): PAN Parks Verification Manual

PAN Parks (2009): As nature intended Best practice examples of wilderness management in NATURA 2000 network. PAN Parks Foundation, 40 pp.

SAARINEN, J. (1998): Wilderness, Tourism Development, and Sustainability: Wilderness Attitudes and Place Ethics. In USDA Forest Service Proceedings P 4.

ŠVAJDA, J. (2002): Attendance of alpine zones of TANAP and possibilities for its regulations. Ochrana prírody Slovenska 2/2002: p. 22. (in Slovak)

ŠVAJDA, J. (2009): Evaluation of Integrated Protected Area Management in Slovak National Parks. MSc. Thesis, University of Klagenfurt, 112 pp.

TAYLOR, D. (2004): Managing the visitor experience within Europe's protected areas. PAN Parks Foundation, 101 pp.

THOMAS, L. & H. MIDLETON (2003): Guidelines for management planning of Protected Areas. IUCN, Gland, pp. 79

VANČURA, V. (2002): PAN Parks as a reality. Ochrana prírody Slovenska. 4, p. 22-23. (in Slovak)

VANČURA, V., Z. KUN & M. VAN DER DONK (2008): PAN Parks Perspectives for a Wilder Europe. International Journal of Wilderness 14 (1): 38-42.

VAN DER DONK, M. (2000): Visitor Management PAN Parks. Framework for a Visitor Management Plan. Thesis. 71 pp.

VOLOŠČUK, I. (1999): The National Parks and Biosphere Reserves in Carpathians – The Last Nature Paradises. ACANAP Tatranská Lomnica, Slovak Republic, 248 pp.

VOLOŠČUK, I., BERKOVÁ, A., PAVLÍK, J., JANČURA, P. & ADMINISTRATION OF TANAP, (2004): Proposal of management plan for TANAP – document for negotiation process.

WATSON,A., L. ALESSA & B. GLASPELL (2003): The Relationship between Traditional Ecological Knowledge, Evolving Cultures, and Wilderness Protection in the Circumpolar North. Conservation Ecology 8 (1): 2-21.

Documents:

Act 543/2002 about nature and landscape conservation

Decree of Ministry of Environment of Slovak Republic n. 4/2011 about SPA Tatry

Decree of Ministry of Environment of Slovak Republic n. 3/2004-5.1 about National checklist of proposed SCIs

Rules for visitors issued by Decree of Regional office of environment in Prešov n. 1/1999

6.2 Internet Resources

Administration of Tatra National Park: http://www.spravatanap.org/

Arolla film: http://www.arollafilm.com/

PAN Parks: http://www.panparks.org/

State Forests of TANAP: http://www.lesytanap.sk/

State Nature Conservancy of Slovak republic: http://www.sopsr.sk/

World Database on Protected Areas: http://www.wdpa.org/

List of Acronyms and Abbreviations

CBD	Convention on Biological Diversity
EU	European Union
IPAM	Integrative Protected Area Management
IUCN	International Union for Conservation of Nature and Natural Resources
NGO	non-governmental organization
NP	national park
PA	protected area
SCI	Site of Community Interest
SPA	Special Protected Area
STDS	Sustainable Tourism Development Strategy
SWOT	strengths, weaknesses, opportunities and threats analysis
TANAP	Tatra National Park
UN	United Nations
UNESCO	United Nations Educational, Scientific and Cultural Organization
WDPA	World Database of Protected Areas
WWF	World Wide Fund for Nature

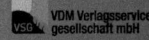